AF824453

First published in 2025
By Kids4kids.com.au publications
Giving a voice to kids to speak to tell their stories and speak to peers everywhere

Text and illustrations copyright 2025 of Anton Mohajeri & Anastasiia Kabanova

All rights reserved. No part of this publication may be reproduced, stored in a retrieval system or transmitted in any form or by any means – electronic, mechanical, photocopying, recording or otherwise – without prior written permission of the publisher and author.

A catalogue record for this book is available at the National Library of Australia

ISBN (Paper back): 978-1-7644009-1-6
ISBN (eBook – Amazon.com): 978-1-7644009-0-9

Where are they?

The mystery of aliens explained for curious minds

By Anton Mohajeri

CHAPTER 1.
WHAT EVEN IS AN ALIEN?

When people say alien, they picture a green creature in a shiny spaceship or a weird tentacle creature, but they are just in the movies and they speak perfect English.

Do you think it's true because I don't think so.

But really, an alien could be anything that isn't from Earth — a tiny germ, a drifting crystal, even a cloud that can think.

It might live in acid, swim in lava, or float above a planet we've never seen.If it isn't born here, it's alien.Because space is so huge, even the smallest chance of life means billions of possibilities.Somewhere out there, something might be asking the same question we are: are we alone?

Scientists prefer the word extraterrestrial — "extra" means beyond, and "terrestrial" means Earth.

It just means "beyond Earth."

They're not looking for monsters or invaders; they're searching for chemistry — for water, gases, and clues that could lead to life.

They use telescopes, rovers, and even listening dishes to hunt for whispers from the stars.

Every signal, every rock sample, is part of the greatest detective story ever told.

Carl Sagan Once Said

"There are more stars in space than grains of sand on every beach on Earth."

If even one in a million stars has a planet with water, that's still millions of possible homes for life.

When you stand outside at night and look up, remember: each star could have worlds of its own, spinning quietly, maybe even wondering about you.

Life on Earth is tougher than we ever imagined.

Microbes live in boiling volcano pools, under ice sheets, and deep in the dark mysterious ocean.

Tiny animals called tardigrades can survive being frozen, dried, and even blasted into space.

So I mean if life can exist in places like that here, why not elsewhere? Maybe the universe is packed with survivors — just not the kind we've learned to recognize yet.

Mars, Europa, Enceladus — these worlds all have the ingredients life loves: water, warmth, and time.

Maybe under the ice, in a hidden ocean, something already swims.

And maybe, somewhere far away, something looks back across the stars and wonders who we are.

That's how all great adventures begin — with curiosity and a question: Where are they?

CHAPTER 2.
OUR STRANGE WONDERFUL WORLD

Earth is the only world we know that drips with water, breathes oxygen, and hums with life.
It sits in just the right spot from the Sun — the Goldilocks Zone, not too hot, not too cold.
If we were closer, oceans would boil; farther, they'd freeze.
Everything about Earth seems designed for comfort, yet it's also wild: lightning, earthquakes, volcanoes, and storms still rage.
Life didn't need peace to begin — it only needed a chance.

Long ago, our planet was molten rock and poisonous gas.

Tiny cells learned to live in that chaos, feeding on sunlight, growing into plants, fish, and finally humans.

That slow miracle shows us that life doesn't start perfect — it builds itself.

So when we look at other planets, we search for the same hints: light, warmth, and patience.

Given enough time, chemistry can become curiosity.

Scientists call those hints biosignatures — gases like oxygen or methane that might come from living things.

When telescopes find such signs, it's like spotting smoke and wondering where the fire is.

A single chemical fingerprint across space could tell us a planet is breathing.

Even extreme worlds tempt imagination. Venus has acid clouds; Titan has lakes of liquid methane. Yet both may hide microscopic guests. Life is clever — it shapes itself to its home, not the other way around.

Understanding Earth isn't just pride; it's practice.
By studying our own fragile miracle, we learn where else miracles might happen.

CHAPTER 3.
THE PLANETS NEXT DOOR

Mars is our red neighbour, dusty and dramatic.
Ancient riverbeds snake across its surface.
Rovers crawl through its sand, hunting for fossils the size of a grain of salt.
If they find even one, it means life is not unique.

Farther out, Europa, a moon of Jupiter, hides an ocean beneath its ice shell. Heat from its core may keep that water liquid.
In the dark, vents could release minerals — food for unseen creatures.

Enceladus, a moon of Saturn, sprays jets of water vapour into space. Spacecraft have flown through those plumes and sniffed organic molecules — the building blocks of life.

Venus once seemed hopeless, but some scientists spotted strange gas high in its clouds — maybe a sign of microbes drifting in the mist.
Hot, yes, but not empty.

Even Pluto, cold and distant, may hide a slushy heart of water beneath its ice.

The lesson is clear: the search for aliens doesn't begin in distant galaxies.
It begins right next door.

CHAPTER 4.
HOW SCIENTISTS HUNT FOR ALIENS

Astrobiologists are detectives of the cosmos.

Their clues aren't footprints or fingerprints — they're wavelengths, chemicals, and signals.

Every telescope is a magnifying glass aimed at forever.

The James Webb Space Telescope peers at exoplanets — worlds orbiting other suns.

By splitting their light into colors, it can see which gases float in their air.

A tiny dip in starlight might reveal an entire atmosphere

On Mars, the rovers Curiosity and Perseverance grind stones and drill cores.

Under microscopes, a pattern of curves could be an ancient microfossil.

Each sample is a postcard from another time.

Back on Earth, giant radio dishes listen for whispers from the stars.

In 1977 one recording screamed a burst of sound — the Wow! Signal — then silence.

Maybe interference, maybe a greeting.

The mystery remains delicious.

To hunt for life is to combine patience with imagination.

It's slow science — but every signal we record feels like the universe answering, keep listening.

CHAPTER 5.
TINY ALIENS FIRST?

Most scientists think the first aliens we'll meet won't have faces or voices.

They'll be microbes: simple, single-celled survivors.

We'll probably find them in ice or dust, not cities

Microbes on Earth live in boiling acid, freezing caves, and radioactive waste.

If they can thrive here, maybe cousins of theirs thrive elsewhere.

They prove that life's first rule is "adapt or disappear."

Finding microbes would be world-changing.

It would mean life wasn’t a one-time accident; it’s a common pattern.

Suddenly, every star would seem friendlier, every planet more possible.

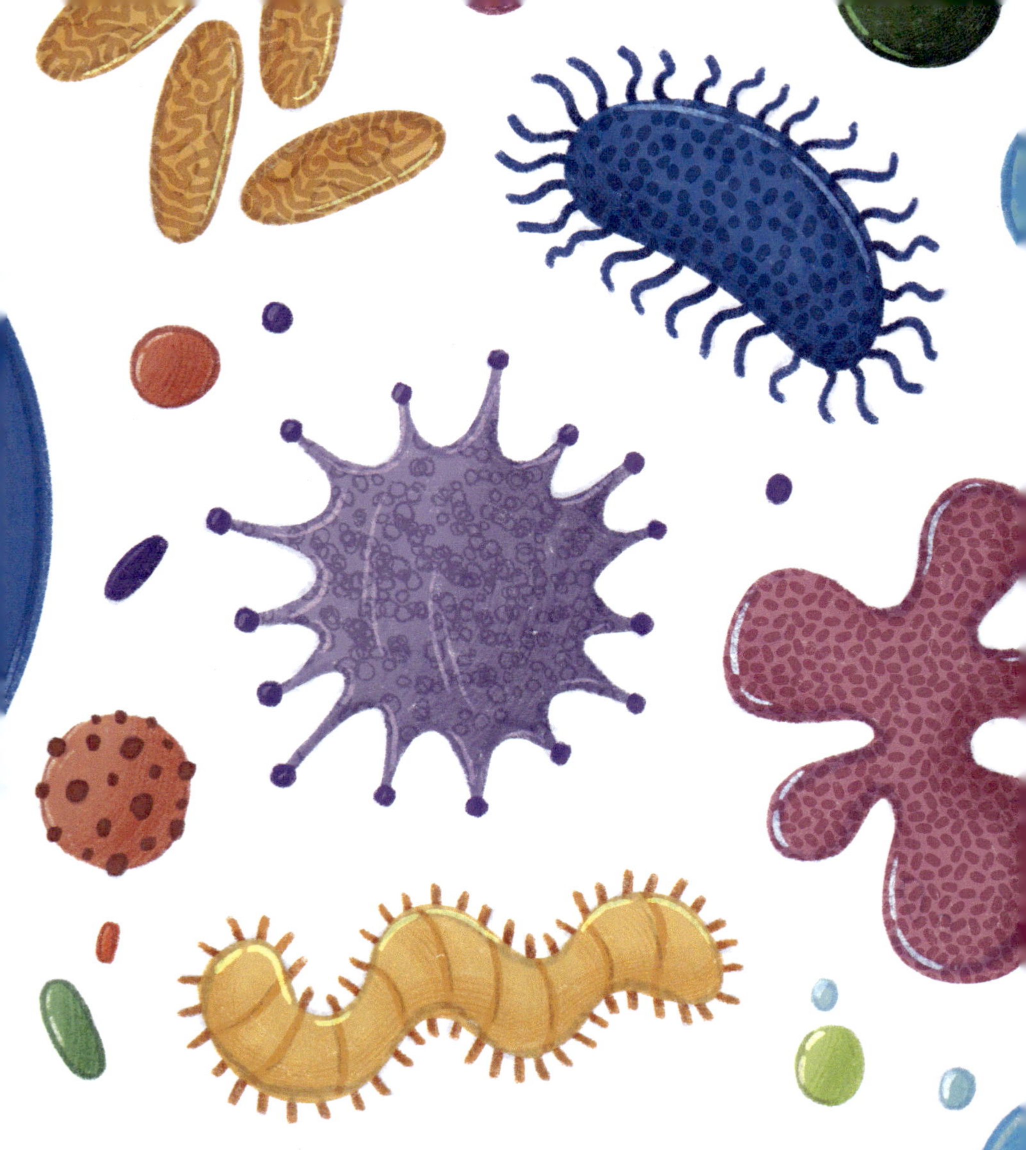

A microscope view of a wriggling alien cell might not look dramatic — but it would be history's biggest discovery.

After that, the real question begins: If microbes are everywhere, where are the minds?

CHAPTER 6.
MESSAGES IN THE STARS

HOW DO YOU SAY HELLO ACROSS SPACE?

Imagine trying to text someone on another planet. No phone, no Wi-Fi, no delivery report. You just send a message that might take thousands of years to arrive… and you might be dead by the time they read it. Still, scientists are trying.
They call it METI — Messaging Extraterrestrial Intelligence. Basically, they're cosmic pen pals.
If we ever say "hello," what language should we use? English? Klingon? Cat? (Cats already act like aliens anyway.)
Scientists think numbers might be best, because math doesn't change. One plus one equals two everywhere — unless you're bad at maths, then it's "two-ish."

Question time:
If you could send one emoji to the universe, what would it be —

🚀, 🐙, 💩, or 🤔?

THE ARECIBO GREETING

In 1974, astronomers used a giant radio dish in Puerto Rico to send a message into space. It was made of dots that formed a human, some DNA, and numbers. Think of it as a super old-school pixel art postcard.

It's flying toward a cluster of stars 25 000 light-years away. That's so far that by the time it gets there, our great-great-great-great-grandkids will probably be communicating by telepathy (or brain-Wi-Fi).

Mini thought experiment:
If aliens reply in another 25 000 years, will anyone even remember what we sent?

VOYAGER'S GOLDEN RECORD

The Voyagers, two space probes launched in 1977, carry actual records — like the kind your grandparents dance to. Each gold-plated disc holds greetings in 55 languages, whale songs, laughter, and even rock and roll.

If aliens ever find it, they might think Earth is a very noisy place full of cheerful mammals.

Question:
If you had to record one sound from Earth for aliens, what would it be — birdsong, waves, a baby laughing, or your little brother singing ?

SHOULD WE WHISPER OR SHOUT?

Some scientists want to keep broadcasting. Others say, "Shhh!" because we don't know who's out there.

What if the first species to answer is a cosmic spammer? "Dear Earthling, you've won 1 billion galactic credits! Click here to collect!"

Still, curiosity wins. The universe is too big to stay quiet forever.

Try this:
Go outside tonight, look at the stars, and whisper "hello."
Congratulations — you've joined the biggest conversation ever.

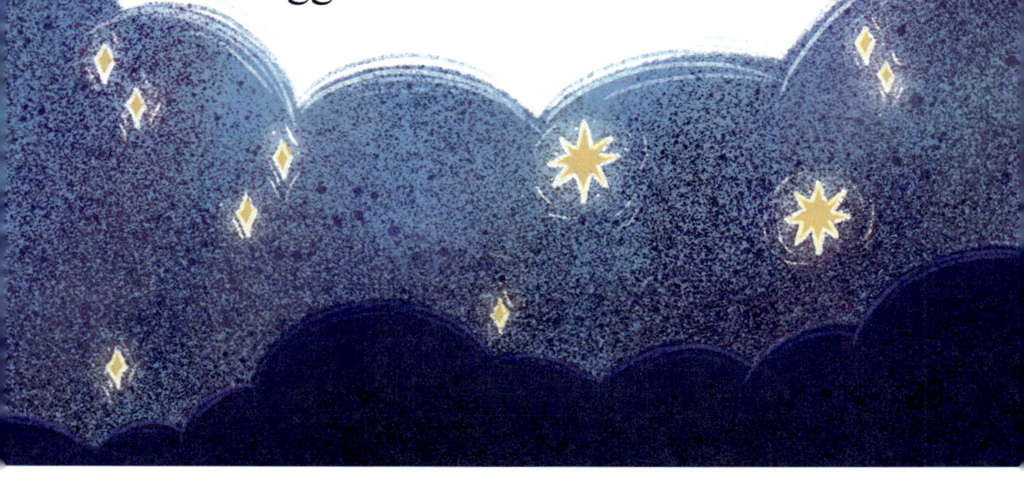

OUR COSMIC INBOX

Every telescope is a giant ear, waiting for a ping, a whisper, or a "Your up?" from space.

So far, silence. But silence isn't nothing — it's A mystery, and mystery's are delicious.

When a signal does come, the first thing we'll do is argue whether it's aliens or the toaster.

And that's okay. Every unanswered message makes the night sky feel alive.

CHAPTER 7.
THE WOW! AND OTHER WHISPERS

THE DAY THE UNIVERSE SHOUTED "WOW!"

In 1977, a telescope in Ohio caught a signal so clear that the scientist wrote "WOW!" next to the data. No one knows where it came from. Maybe aliens, maybe a comet, maybe a space microwave.

It never happened again — like getting one text that says "hi" and then ghosted for 50 years.

Question:
If you got a message from aliens that only said "WOW!," would you reply "WOW back" or "Who are you?"

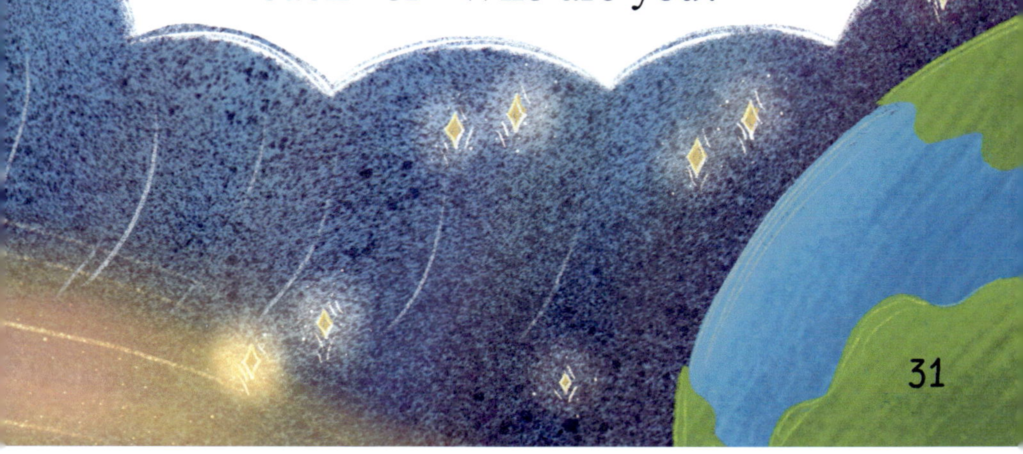

Since then, astronomers have aimed radio dishes at the same spot, hoping for another "wow."

Nothing. Silence can be louder than noise sometimes.

But we've found other odd bursts — "Fast Radio Bursts." They last less than a blink and come from galaxies billions of light-years away.

It's like cosmic popcorn: random, loud, and probably not intentional.
But who knows?
Maybe someone's trying Morse code and we're missing the pattern.

WHEN LUNCH BREAKS CAUSE DISCOVERIES

One time, scientists thought they'd found a repeating alien signal.

They tracked it down to… the observatory microwave. Someone kept opening the door mid-heat.

That's science for you: sometimes the mystery of the universe is just leftovers.

Mini-Challenge:
Next time something weird happens, check your kitchen before declaring "ALIENS!"

PATTERNS AND PATIENCE

Scientists record everything — even static.

They write computer programs that scan for repeating patterns, like heartbeat rhythms or musical scales.

Imagine if we got a perfect tune across space — a cosmic ringtone!

For now, the signals keep teasing us — one second there, next second gone.

That’s the beauty of science: it keeps us listening even when the universe plays hard to get.

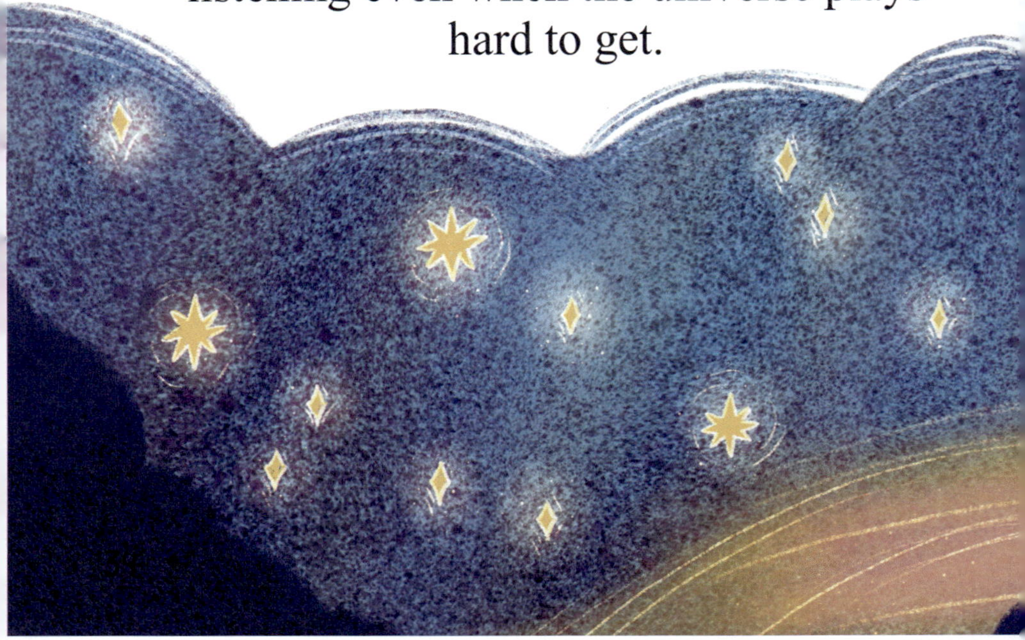

THE LESSON OF THE WOW!

Maybe “WOW!” wasn’t a message, but it changed how we listen.

It taught us that discovery can start with one excited word in the margin.

So next time you feel amazed by something — a frog, a comet, a piece of toast that looks like Einstein — remember, that’s your own Wow signal.

CHAPTER 8.
THE FIRST CONTACT QUESTION

WHO GETS TO ANSWER THE PHONE?

Imagine aliens finally call. Who picks up? NASA? The president? A random teenager on TikTok yelling, “Hey bestie, welcome to Earth”?

No one’s sure.

There’s no rulebook called “How to Chat with Extraterrestrials for Dummies.”

Question:
If the message came through your TV right now, what’s the first thing you’d say?

EARTH’S DEBATE CLUB

Some scientists think we should hold a worldwide meeting before answering.

Others say, “Too late, they already know we’re noisy.”

There’s even an official plan at the United Nations… which mostly says “We’ll talk about it later.”

Classic bureaucracy.

If aliens are listening, they probably think Earth’s main language is arguing.

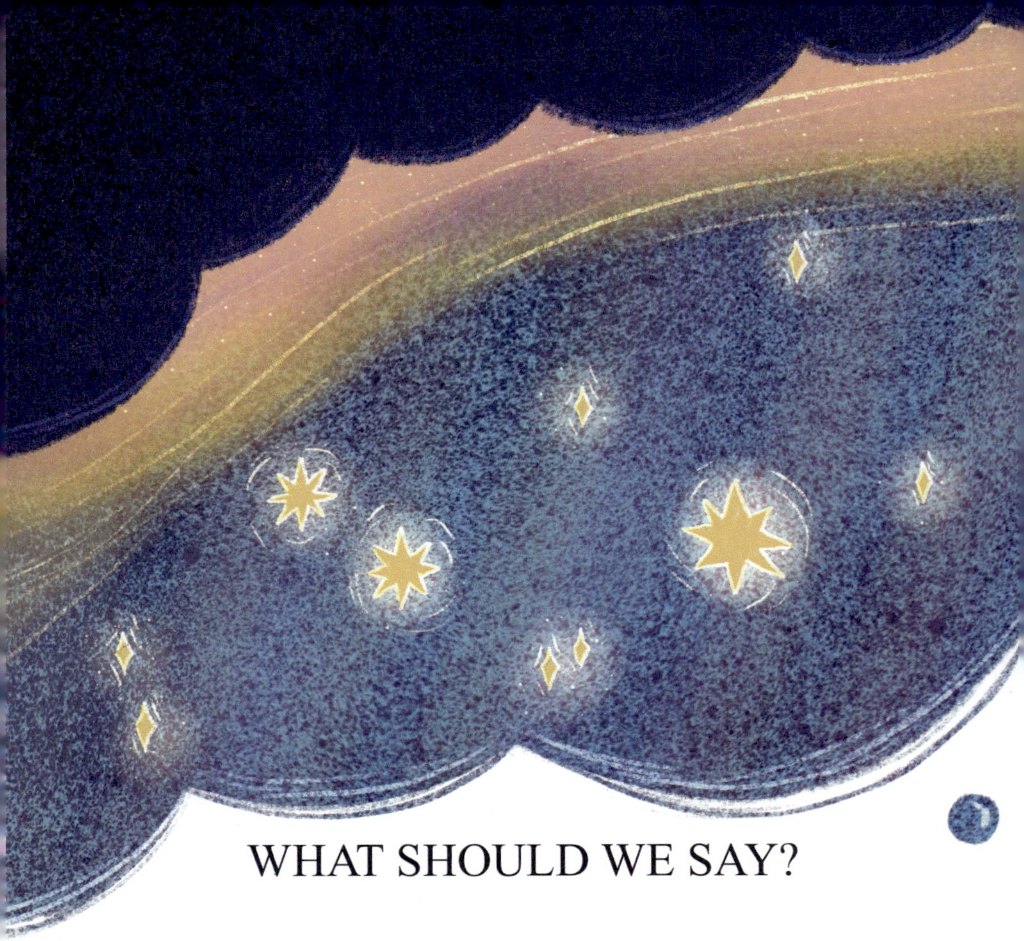

WHAT SHOULD WE SAY?

Maybe start simple:
"Hi, we're a bunch of smartish monkeys. Sorry about the noise. Do you like water?"

Or maybe send a math problem — if they solve it, we'll know they're clever.

Whatever we say, it should be kind, honest, and not include our exact address (lesson learned from the Internet).

SCARY OR EXCITING?

Some worry first contact could be dangerous.

What if they're conquerors, not friends?

History tells us that first meetings between civilizations can get messy.

But others say fear shouldn't stop curiosity.

If everyone stayed home scared, we'd still think the Earth is flat and the Moon is made of cheese.

THE HELLO THAT CHANGES EVERYTHING

Maybe the first message won't be words at all — just a pulse of light saying, We're here.

And maybe our reply will be the same.

When that day comes, I hope we're ready — polite, funny, and wearing clean shirts.

Because first impressions matter, even when they travel at the speed of light.

CHAPTER 9.
ALIEN PLANETS EVERYWHERE

THE GREAT PLANET HUNT

Every night, telescopes stare at the sky like giant unblinking eyes, searching for tiny dips of light — that's how they spot planets going in front of their stars. We've already found more than 5 000! That's like discovering a whole new set of neighbours and realising the galaxy's a crowded apartment building.

Most of them are too far to visit, but every time one is found, Earth feels a bit less lonely.

Question:
If you could name a new planet, what would you call it — "Planet Awesome", "Bob", or "Chocolate 2"?

WEIRD WORLDS

Some planets rain glass. Others have winds faster than sound. One, called 55 Cancri e, is made mostly of diamond — talk about bling!

Another one is so close to its sun that it glows red and smells like burning metal.

Not ideal for a holiday, unless you're into sauna death traps.

Imagine:
If aliens visited our planet, what would they say about Earth? Probably, "Lovely oceans, but too many reality shows."

THE TRAPPIST-1 FAMILY

Seven rocky planets orbit one tiny star called TRAPPIST-1.

Scientists think at least three could have liquid water.

That's like a solar system with multiple chances for life — a cosmic multiplex!

If any microbes live there, they might already be staring up at their star, wondering if someone's staring back.

Try this thought:
If you could travel there but it took 40 years, would you go?

SUPER-EARTHS AND MINI-NEPTUNES

Some exoplanets are “super-Earths” — bigger, heavier, and with stronger gravity.

You’d weigh twice as much!

Others are “mini-Neptunes” — smaller gas giants that could have floating clouds where aliens swim instead of walk.

It’s like every possible planet idea ever got built somewhere out there.

INFINITE HOMES, INFINITE HOPE

If there are trillions of planets, even a tiny chance of life means the universe must be buzzing somewhere.

So next time you look at the night sky, remember: those dots aren't decorations — they're addresses.

CHAPTER 10. THE HABITABLE ZONE ISN'T ENOUGH

JUST RIGHT… OR JUST WRONG?

We call it the Goldilocks Zone — not too hot, not too cold, perfect for water.

But even if a planet's comfy temperature-wise, it could still be a nightmare: acid rain, exploding volcanoes, or a sky of poison gas.

The universe isn't a five-star hotel.

It's more like a mystery buffet — you never know what you're biting into.

LIFE HIDES IN WEIRD PLACES

On Earth, microbes live in volcanoes, under ice, and inside nuclear reactors.

So why wouldn't aliens do the same? They might live miles underground, in darkness, slurping minerals like soup.

Question:
Would you rather explore a planet of lava or a planet of ice? Choose wisely — your boots depend on it.

TITAN'S COLD BATHTUB

Saturn's moon Titan has lakes of liquid methane — basically, space gasoline.

If aliens live there, they'd drink methane instead of water and burp fire (pretty metal).

Maybe they move slowly, frozen and calm, thinking thoughts a million times slower than ours.

REDEFINING “ALIVE”

We keep searching for Earth-like life, but maybe that’s too narrow.

Maybe life doesn’t even need cells — maybe it’s plasma storms that think or magnetic fields that feel.

The universe could be full of minds too different for us to recognise.

Big question:
What if aliens already live in the Sun — and we just can’t see them because we’d burn?

LIFE 1.5: THE UPGRADE

So “habitable” doesn’t mean “like us.”
It means “possible.”

And possible is the most exciting word in science.

CHAPTER 11.
THE GREAT FILTER

THE SILENT GALAXY MYSTERY

If the galaxy has so many worlds, where is everyone?

That's the Fermi Paradox: "Where are they all hiding?"

Maybe we're early. Maybe they've gone extinct. Maybe we're the weird ones.

Joking thought:
Maybe aliens did come, saw the Internet, and left quietly.

THE FILTER THEORY

The Great Filter means there's a step in evolution most civilizations don't survive.

It could be asteroid strikes, climate collapse, or inventing social media before emotional maturity.

If that's true, life is fragile — a one-in-a-billion lottery ticket that keeps tearing itself up.

BEFORE OR AFTER?

If the filter is ahead of us, yikes. We haven't passed the test yet.

If it's behind us, congratulations — we're cosmic champions.

Either way, it means survival isn't guaranteed.

Question:
What do you think the biggest danger to humanity is — asteroids, robots, or just being silly?

EXTINCTION LESSONS

The dinosaurs had a bad day 66 million years ago.

Their asteroid wiped out 75% of life, but also made room for mammals — and eventually, you.

Sometimes endings are really awkward beginnings.

THE QUIET HOPE

Maybe it's not a graveyard out there — just a library of patient civilizations waiting for us to grow up.

Maybe silence is politeness on a universal scale.

CHAPTER 12.
ZOO THEORY AND THE WATCHERS

ARE WE BEING OBSERVED?

Some scientists wonder if Earth is like an exhibit in a cosmic zoo.

Aliens might be watching, thinking, “Don’t tap the atmosphere, the humans get jumpy.”

If that’s true, we’re the pandas — cute, endangered, and bad at breeding cooperation.

THE PRIME DIRECTIVE

If aliens follow a "no interference" rule, they won't reveal themselves until we're ready.

Like teachers waiting for the class to stop throwing paper airplanes.

Maybe UFOs are just progress reports. "Still fighting. Still weird. Needs improvement."

TOURIST ALIENS?

Imagine alien tourists with space cameras taking selfies by the Great Barrier Reef: "Visited primitive planet #3. Locals obsessed with coffee and arguing."

Then they zoom home to post, "10/10 atmosphere, would orbit again."

Question:
If you could ask the watchers one question, what would it be?

GRADUATING FROM THE ZOO

Maybe the final test to join the galactic club isn't technology — it's kindness.

Once we prove we can share a planet without wrecking it, the invisible fence opens.

WHEN THE CURTAIN LIFTS

One day, we might look up and realise we've been surrounded all along.

Maybe they'll appear, not as conquerors, but as neighbours saying, "We were waiting for you to stop fighting over parking spots."

CHAPTER 13.
UFOS AND UAPS: SORTING FACT FROM FANCY

THINGS IN THE SKY

Sometimes people see bright lights zipping around the sky and shout, “UFO!” Then it turns out to be a weather balloon, a jet, or Uncle Bob’s new drone.

But sometimes, the sky just… refuses to explain itself.

That’s why scientists now say “UAP” — Unidentified Aerial Phenomenon — because it sounds smarter and avoids arguments about little green pilots.

Question:
If you filmed something weird in the sky, would you post it online, call NASA, or run for snacks?

GOVERNMENTS GO PUBLIC

Even governments have admitted: yes, pilots have seen things they couldn't identify.

In 2020, the Pentagon released real videos. One object zipped like a pinball on caffeine.

Did aliens visit? Maybe. Or maybe the cameras glitched.

Science's job isn't to believe or deny — it's to investigate until it's boring.

OPTICAL TRICKSTERS

Sometimes our brains make patterns out of nonsense.

Clouds look like elephants, stars like faces, flying bags like saucers.

That’s called pareidolia — the art of seeing what isn’t there.

It’s our brain’s way of saying, “I’ll fill in the blanks, thanks.”

REAL MYSTERIES ARE RARER

Out of thousands of reports, only a handful stay mysterious.

But even one "unexplainable" sighting keeps hope alive — and telescope funding healthy.

Scientists secretly love mysteries. They make the universe spicy

WONDER WITHOUT PANIC

So when you see something odd, don't scream "ALIENS!" or roll your eyes.

Say, "That's interesting." That's the magic scientist phrase.

Wonder and skepticism aren't enemies — they're a power couple.

CHAPTER 14.
ANCIENT VISITORS OR ANCIENT IMAGINATION?

THE PYRAMIDS PROBLEM

TV shows say aliens built Egypt's pyramids.

Archaeologists say: humans did — with math, muscles, and bread.

Aliens didn't show up with blueprints and laser levels.

They weren't needed. Ancient engineers were geniuses with patience.

Question:
Do you think aliens would waste time stacking rocks when they could just float above them?

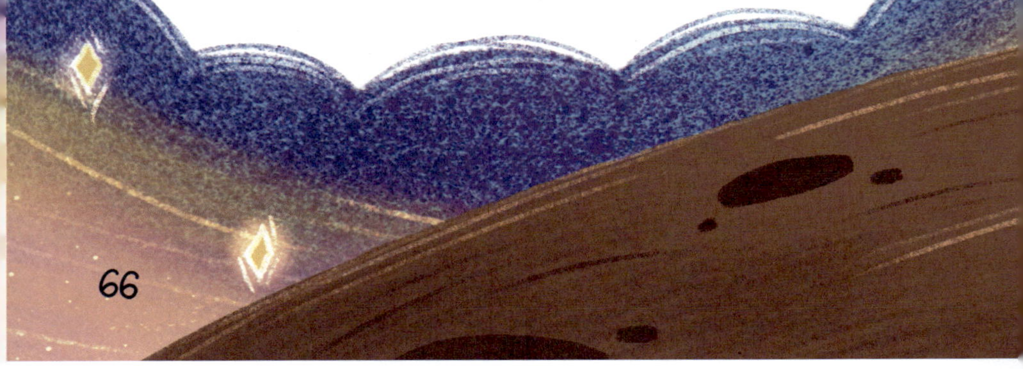

SKY GODS AND LEGENDS

Old stories describe glowing discs and fiery chariots.

Maybe those were meteor showers, comets, or volcanoes erupting.

When you don't have science words, you use poetry.

Mythology was the first science — beautiful guesses about confusing stuff.

WHY PEOPLE LIKE ALIEN BUILDERS

Because it's exciting! It makes the past magical and the ordinary cosmic.

But it also sneaks in a sad idea: that humans couldn't have been smart enough.

We were. We still are.

It's just easier to say, "aliens did it" than to study 20 years of archaeology notes.

TESTING THE STORY

Science doesn't kill wonder; it cleans it.

It says, "Cool theory. Show me the evidence."

No proof? Then it's imagination — and imagination is awesome too, just in a different category.

CREDIT WHERE IT'S DUE

Ancient humans built calendars, pyramids, and cities — all with stone tools and stubbornness.

They deserve the applause.

Aliens can build their own pyramids, thank you very much.

CHAPTER 15.
WHAT WOULD THEY LOOK LIKE

SHAPE SHIFTERS

Forget green stick figures.
Aliens could look like bubbles, worms, light, or trees.

Evolution designs for survival, not fashion.

If you live on a windy world, you'll be short and strong; in low gravity, tall and floaty.

THE AIR WE BREATHE (OR DON'T)

Earthlings need oxygen.

Methane-breathers might have metal blood, or glow to find food.

Sulfur-loving microbes could smell like rotten eggs but call it "perfume."

Beauty, it turns out, is atmosphere-dependent.

Question:
Would you shake hands with an alien if it had three tentacles and smelled like cheese?

SEEING WITHOUT EYES

Who needs eyes when you can sense magnetism?

Some creatures on Earth already do.

Aliens might "see" heat or sound waves instead of light.
Maybe they dream in smells. Imagine that! A scented movie.

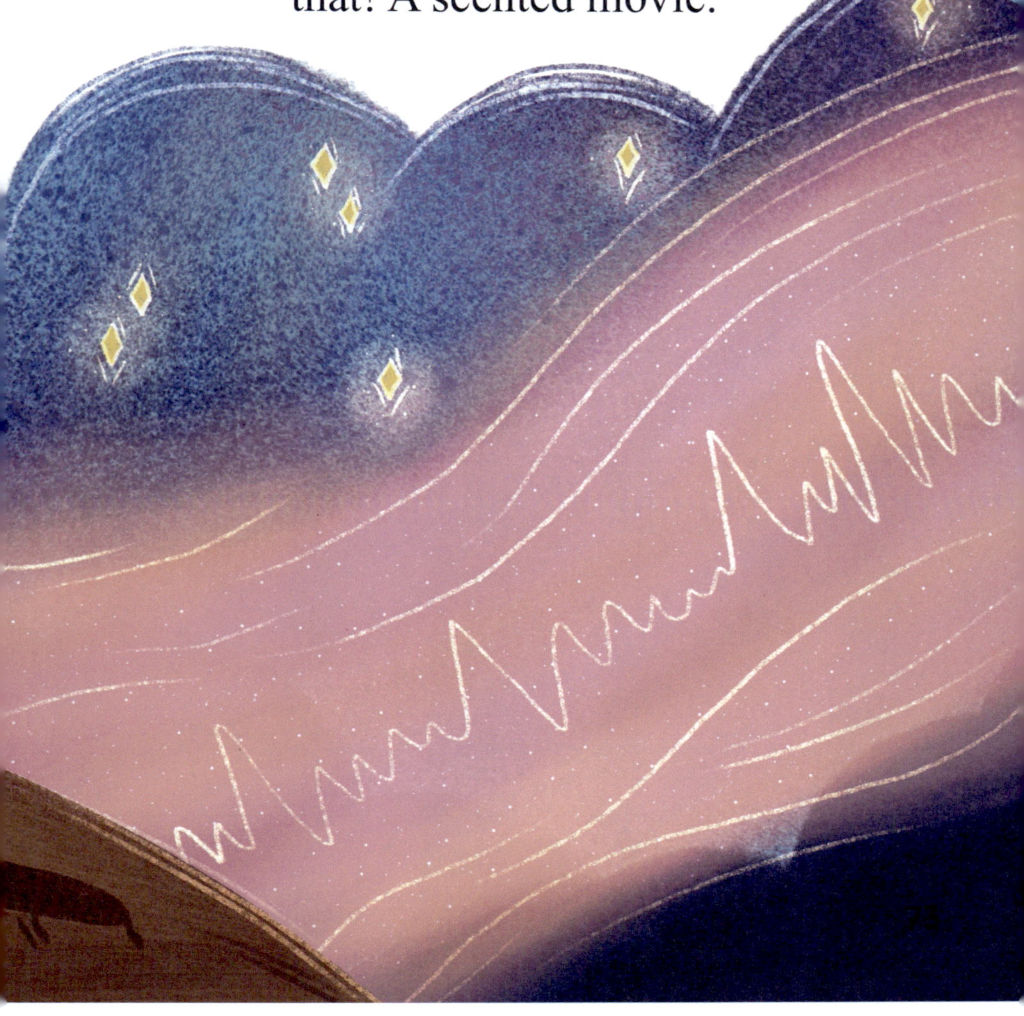

THINKING FORESTS

Brains aren't the only way to think.
A forest could send messages through its roots.

A cloud of plasma might store memory in electric ripples.

"Life" might not even need skin.
So yes, your next houseplant might secretly be judging you.

THE MIRROR TEST

When we finally meet them, they'll make us redefine "alive."

Maybe they'll look strange.

But they might look at us and whisper, "Now that's weird."

CHAPTER 16.
COULD WE BE THE ALIENS?

COSMIC MAIL DELIVERY

Panspermia: life traveling by rocks.

When asteroids hit planets, they fling debris into space — sometimes with microbes hitching a ride.Earth might have been seeded by space bacteria from Mars.

That means we could be Martians in disguise. Surprise!

METEORITE CLUES

Some meteorites from Mars have weird tiny shapes inside them — maybe fossilized cells, maybe not.

Either way, it's proof that planets swap more than just dust.

We've been cosmic pen pals for billions of years.

COMET CHEMISTRY

Comets carry amino acids — the Lego bricks of life.

When they crash into planets, they deliver ingredients like a cosmic grocery drop.

Whoever planned that delivery system deserves five stars.

FAMILY TREE OF THE COSMOS

If all life came from the same spark, then the universe is one big family reunion.

You, me, tardigrades, and any aliens out there — distant cousins at the table.

Please pass the stardust.

THE BEAUTIFUL MAYBE

Whether life started here or there doesn't matter as much as this:

It started. It survived.

We're proof the universe doesn't like being quiet.

CHAPTER 17.
THE FUTURE OF EXPLORATION

ROBOTS TAKE THE RISK

Rovers don't need bathrooms or snacks.

They just trundle along, taking selfies on Mars.

Perseverance, Curiosity, and soon, the Europa Clipper — metal heroes doing science miles from home.

TINY SHIPS, BIG DREAMS

Scientists are building mini-probes with light sails.

They'll be pushed by lasers, racing to other stars faster than any rocket.

Imagine: postcards from another sun in your lifetime.

HUMANS JOIN THE PARTY

We'll build moon bases first, then Mars camps.

We'll grow lettuce in domes, fix robots, and argue over chores — just like Earth, but redder.

Question:
Would you rather be the first Martian baker or the first Martian teacher?

SCI-FI BECOMES DIY

3-D printers might make houses from lunar dust.

AI will guide spaceships while humans nap.

Spacesuits could recycle sweat into water (ew, but smart).

The future's weird and wonderful — like all good ideas.

THE EXPLORER'S RULE

Every footprint we leave should be kind.

Exploration without care is invasion.

So when we reach new worlds, let's remember: curiosity should never crush what it finds.

CHAPTER 18.
IF WE FOUND THEM TOMORROW

THE NEWS FLASH

"BREAKING: Scientists discover alien microbes!"

Everyone on Earth forgets their passwords for a moment.

Markets dip, memes explode, philosophers faint.Then we all look up. The universe just got personal.

THE REACTION OLYMPICS

Some people would celebrate.

Some would panic.

Some would start printing “Welcome, aliens!” T-shirts.

Religions might argue.

Kids would make the best YouTube videos explaining it all.

THE BIG QUESTIONS

Do they have families? Art? TikTok?

Can we be friends? Trade? Should we?

The first discovery wouldn't just be about them — it'd be about us.

It would show who we really are when faced with the unknown.

GOVERNMENT SECRETS?

Would officials hide the news "to prevent panic"? Maybe for five minutes.

Then someone would tweet a blurry photo, and the world would lose its collective mind.

Truth moves faster than rockets.

THE MORNING AFTER

The sun would still rise. Coffee would still smell amazing.

But somewhere deep inside, every human would feel taller — part of a bigger story.

For once, Earth would be united by curiosity instead of competition.

CHAPTER 19.
THE UNIVERSE AS A LIVING THING

EARTH THE ORGANISM

Some scientists think Earth behaves like a giant living creature.

Forests breathe carbon, oceans pulse like blood, and storms act like sneezes.

It's called the Gaia Hypothesis — fancy name, cool thought.

COSMIC BODY

If Earth can act alive, what about the universe?

Galaxies could be its organs; stars, its cells.

Dark matter might be connective tissue, holding it together.

We might live inside a thinking giant — call it "Bob the Cosmos."

CONSCIOUS OR COINCIDENCE?

It’s not proven, just poetic science.

But maybe awareness isn’t limited to brains.

Maybe existence itself is a kind of thought.

That would make every atom a spark of attention.

YOU ARE STARDUST (LITERALLY)

The iron in your blood came from exploded stars.

So did your bones, your teeth, your phone, and your cat.

You're not just in the universe — you're of it.

You're the universe noticing itself in the mirror.

COSMIC CHILL

If the universe is alive, it's the most relaxed being ever.

It takes billions of years to blink.

We're just brief thoughts passing through its imagination.

CHAPTER 20.
ARE WE READY?

Before meeting aliens, we should probably stop fighting about parking spots and Wi-Fi passwords.
No one invites rude guests to the Galactic Club. Maybe the test isn't technology — it's empathy.

We've polluted oceans and melted ice caps.
If aliens visit, they might bring clipboards: "So, how's your planetary management going?"
We'd better have better answers than, "Oops."

KINDNESS: THE NEW ROCKET FUEL

Maybe kindness is the universe's ultimate intelligence test.

If we can treat each other well, maybe we'll treat aliens well too.

The first interstellar handshake should come with good manners.

THE GALACTIC INVITATION

When we finally learn to cooperate, maybe we'll hear a new signal — a cosmic ding!

"Congratulations, Earth. You're ready. Please bring snacks."

Question:
What snack would you bring to meet aliens? (Pro tip: not chili.)

Until then, we keep wondering, exploring, and laughing. Because that's what intelligent life does. We look up at the stars and ask questions.

And somewhere, maybe, someone looks back — smiling.